WALTER DUNCAN BELLINGRATH
1869-1955

BESSIE MORSE BELLINGRATH
1878-1943

Founders of Bellingrath Gardens

Born in Atlanta, Georgia, Walter D. Bellingrath began his career at the age of 17 as a railroad station agent and telegraph operator for the Louisville and Nashville Railroad in the small Alabama town of Castleberry. At the time of his death at 86, he had become one of the South's great leaders in business, church, educational and civic affairs. He moved to Mobile in 1903 and founded the Coca-Cola Bottling Company there. He later married Bessie Mae Morse of Mobile, whose tireless energy and love of beauty played a predominant part in the masterpiece of their life's work: the creation of Bellingrath Gardens and the establishment of The Bessie Morse Bellingrath Collection of antique furniture, priceless silver, rare porcelains and fine china.

Bellingrath Gardens

There are a few places in the world where magic is still to be found—the magic of beauty so poignant that it fills the heart with awe. Such a place is Bellingrath Gardens, on the Isle-aux-Oies River, near Mobile, Alabama.

Visitors who enter the wrought-iron gates of this vast and lovely estate will soon realize they have come upon a veritable wonderland. Overhead, from branches of aged oaks, trail graceful wisps of Spanish moss. Along the sun-dappled paths the flame of azaleas or the blush of hydrangeas excites the eye with loveliness. An early impression is that of the harmony with which man and nature have combined efforts to produce a magnificent garden. The trail that winds so casually through the whispering pines is frequently enhanced by an artful vista that delightfully blends with the natural surroundings.

One of the happy charms of Bellingrath Gardens is that beauty knows no season here. Mr. Bellingrath was fond of comparing his beloved "Charm Spot of the Deep South" to a lovely lady with 52 gowns, one for each week of the year. The metaphor suggests the constantly changing beauty of the Gardens, ever enchanting, never the same.

From late winter's spectacular azaleas, through springtime's dogwood, roses and hydrangeas, through summer's brilliant foliage plants and flowers, to autumn's exquisite camellias, the pageant moves on, perfumed by Confederate jasmine and sweet olive, sheltered by oak, magnolia, and a variety of other trees.

Comments from the many thousands of visitors who tour the Gardens each year are extravagant with praise. Whatever facet of this many-splendored place appeals most to the individual visitor, all share a common experience: the quiet, peaceful and tranquil atmosphere gives one a feeling of reverence and of closer communion with the one Creator of all beauty.

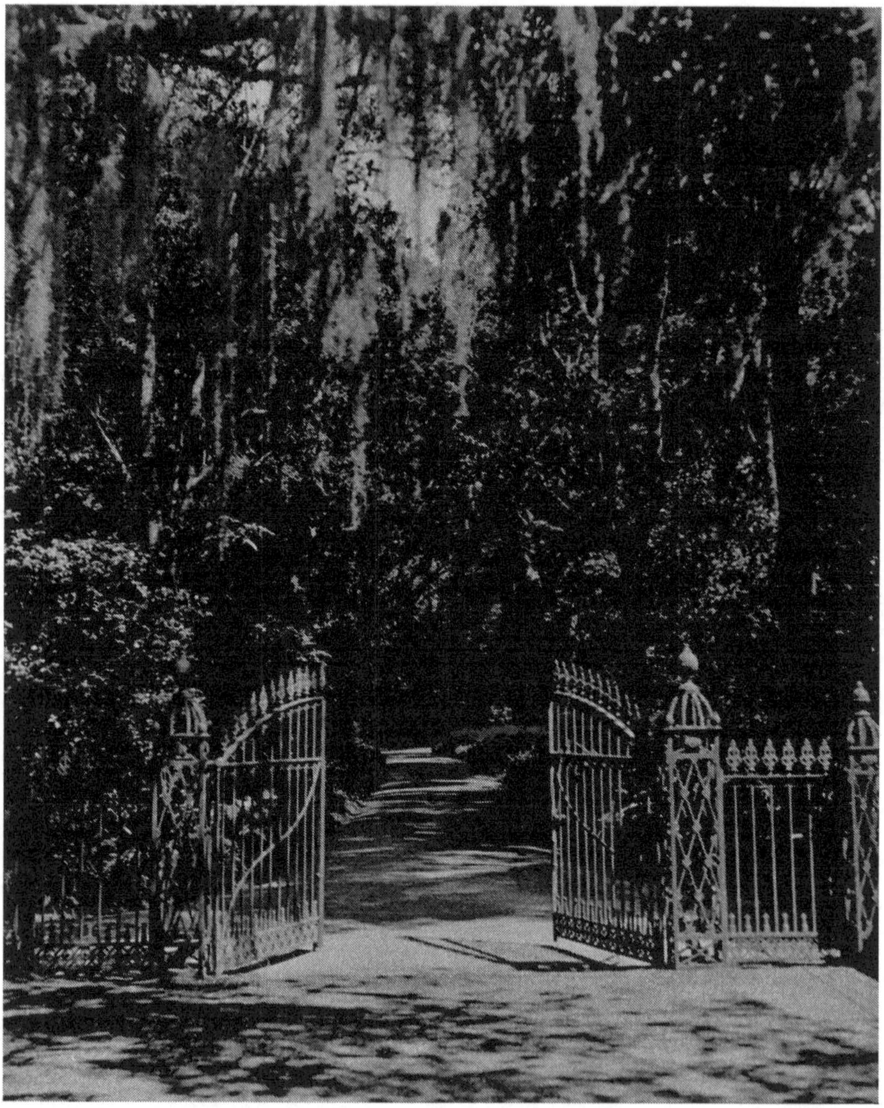

Wrought-iron gates open into a land of enchantment. A bed of scarlet salvia may be seen in the background.

A grassy vista along the entrance path is bordered by hyacinths and azaleas.

Azaleas in full bloom tower overhead to meet trailing Spanish moss.

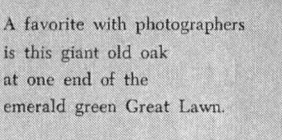

A favorite with photographers is this giant old oak at one end of the emerald green Great Lawn.

White Iveryana azaleas border the path as it nears the center of the Gardens.

A view across Mirror Lake from the top of the fern-lined Rockery.

Cineraria and Easter lilies are combined with azaleas to make this colorful display at the center of the Gardens.

One of Mr. Bellingrath's favorite poems is immortalized in bronze on this plaque.

GOD'S GARDEN

THE LORD GOD PLANTED A GARDEN
IN THE FIRST WHITE DAYS OF THE WORLD
AND SET THERE AN ANGEL WARDEN
IN A GARMENT OF LIGHT ENFURLED

SO NEAR TO THE PEACE OF HEAVEN
THE HAWK MIGHT NEST WITH THE WREN
FOR THERE IN THE COOL OF THE EVE'N
GOD WALKED WITH THE FIRST OF MEN

AND I DREAM THAT THESE GARDEN CLOSES
WITH THEIR SHADE AND THEIR SUN-FLECKED SOD
AND THEIR LILIES AND BOWERS OF ROSES
WERE LAID BY THE HAND OF GOD

THE KISS OF THE SUN FOR PARDON
THE SONG OF THE BIRDS FOR MIRTH
ONE IS NEARER GOD'S HEART IN A GARDEN
THAN ANYWHERE ELSE ON EARTH

DOROTHY FRANCES GURNEY

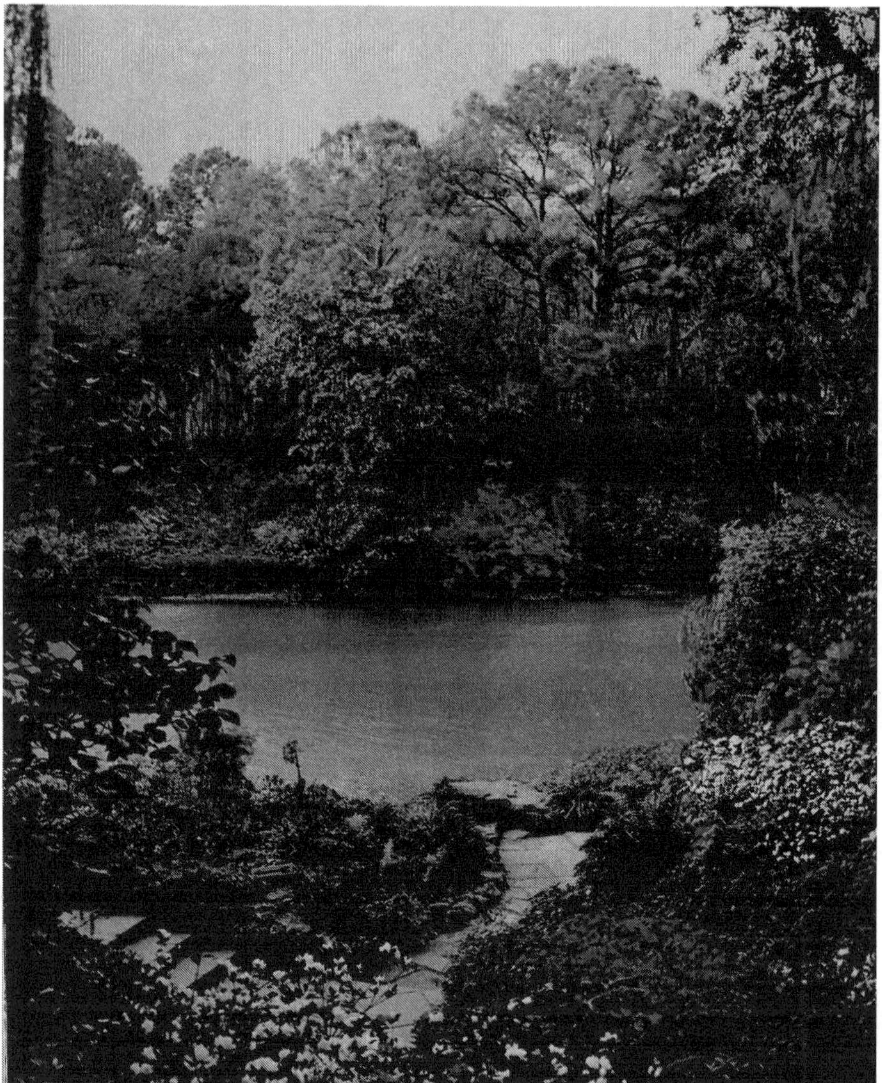

Foliage plants and
semi-tropical shrubs make a
summer border for the Rockery.

White swans sail majestically on
Mirror Lake, which reflects a glossy
magnolia and the red and
yellow of salvia and day lilies.

Descending the flagstone steps
toward the lake, this breathtaking
view comes into sight.

A flagstone path beside Mirror Lake is lined with azaleas and red berried ardisia. The flagstones once paved sidewalks of Historic Old Mobile.

This rustic bridge across Mirror Lake leads to a woodland pathway through towering azaleas.

A view from the far side of the lake. Spanish moss is a
member of the pineapple family, and lives largely on air.

Salvia provides rich color in summer months. The charm of the natural woodlands is retained
In the distance is the Rockery. throughout the Gardens.

Near the
Camellia Arboretum
the iron lace work of
the summer house
frames a lovely view
through the
stately pine trees.

Tulips are colorful
against a bed
of pine straw in
this scene near
the summer house.

Azaleas range in color from white to deepest wine red. These plants are of the Indica (Indian) variety. The Japanese Kurume azaleas are much smaller in bush and blossom. The berried plant, left foreground, is nandina, or Heavenly Bamboo.

The guest house and garage
is located near The Bellingrath Home.

Colorful caladiums are a summer favorite near the Home.

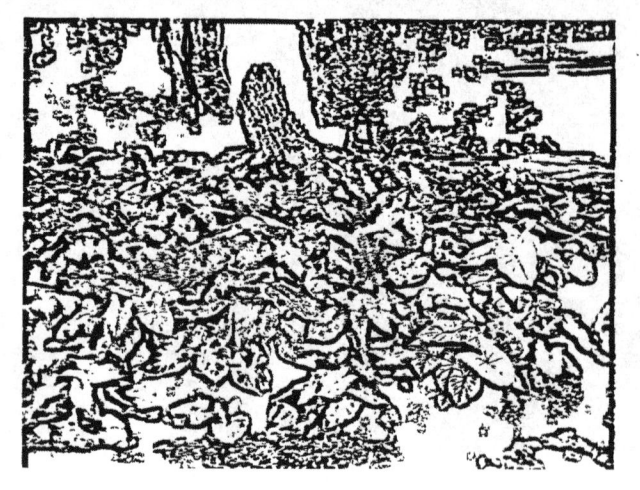

An ancient oak stands in the center of the Gardens . . .

. . . while squirrels and birds play
beneath its spreading branches.

The patio of The Bellingrath Home features seasonal flowers at the height of their beauty. Here are azaleas, Easter lilies and pink hydrangeas.

The Bellingrath Home

Few residences anywhere combine the charm of design and construction, the richness of furnishings and the rarity of art objects that are to be found in The Bellingrath Home.

The furniture, the old English silver, the fabulous collection of china and rare porcelain are known as The Bessie Morse Bellingrath Collection in memory of Mrs. Bellingrath who painstakingly gathered these priceless objects from all over the world.

The house itself, built in 1935, is of handmade brick and wrought-iron lace work, all over a century old. In the words of the late George B. Rogers, the architect who designed it, the house is "a mingling of the French, English and Mediterranean influences, while the interior represents a blend of decor embracing chiefly the English renaissance and colonial America." As handsome as is the exterior of the Home, visitors are still not prepared for the magnificence to be found within.

Here is a comprehensive collection of antique furniture, including French and English pieces that reflect both the Victorian and latter-day French influences. Complementing the furniture are rare 18th and 19th Century pieces of Meissen (Dresden), Sevres and English porcelains. Four different sets of 22-carat gold overlay service plates, one set painted and signed by Angelica Kaufman (c. 1741-1807), are to be seen as well as nine complete dinner services and an exquisite collection of antique silver.

After Mrs. Bellingrath died in 1943, Mr. Bellingrath continued to live in the Home in the center of his beloved Gardens until the time of his death in 1955. It was then opened to the public, according to his wishes, by the non-profit charitable foundation which he had established in 1950.

The drawing room of The Bellingrath Home includes interesting period furnishings.

Note the French porcelain clock and urns by Jacob Petit, c. 1790, on the Adam type mantel. The gold leaf mirror is Louis XVI.

This 18th Century Dresden urn is resplendent in coloring and detail.

The English Chippendale dining table and chairs were once owned by Sir Thomas Lipton. The rare centerpiece is of Meissen porcelain and ormolu.

In the upstairs hall is a hand-carved rosewood console table made for Princess Louise, grandmother of Kaiser Wilhelm II.

This guest bedroom indicates the exquisite detail in which Mrs. Bellingrath had furnished the Home.

This family group in French bisque, a flawless example of the potter's art, is one of several hundred pieces of bisque and porcelain in the Home.

French porcelain urn
(Sevres) c. 1786.

This hand-carved four-poster bed is found in the Purple Room.

Porcelain group (Meissen) Eighteenth Century.

A corner of the upstairs "Morning Room".

Mrs. Bellingrath's bedroom in soft pink with touches of blue has a hand-carved bed made by Mallard of Louisiana, c. 1838—

Mr. Bellingrath's bedroom prominently displays this Jacobean oak desk inlaid in ivory and mother-of-pearl.

This lovely portrait bust
of English bisque by Copeland
has a life-like grace
and softness of contour.

This small dining-room
has many interesting
pieces from
various countries.

A portion of the
China Collection.

The Crystal and Silver Room.

The Bottle Collection is
colorful and unusual.

The fabulous collection of old Georgian,
Sheffield and Sterling silver
includes many articles from
the estate of the Earl of Tankerville,
which dates from the 11th Century.

The Alabama marble on the
kitchen table compares with
Italian marble in color and durability.

In the porch dining-room hang portraits of Generals Washington, Lee and Jackson done in needle point and petit point.

Visitors are invariably impressed by this table with Honduran mahogany top, Chinese blackwood apron and pedestal base, all hand-carved in a popular Jade pattern.

A view of the patio from above shows a summer scene of foliage plants around a fountain.

And a winter view of the patio finds Easter lilies, poinsettias and chrysanthemums blooming.

After reveling in the works of the finest craftsmen of Europe and America, visitors return to the beauty of the out-of-doors in Bellingrath Gardens. After seeing so much, it is difficult to believe that still further surprises in loveliness lie ahead, but they are soon apparent as visitors stroll along the Isle-aux-Oies River, by the Grotto, past the Great Lawn and into the Rose Garden.

Steps descend from front of Bellingrath Home to the Isle-aux-Oies River.

Red-berried ardisia and green ferns fill the urns on this terrace.

A flagstone walk leads along the riverside.

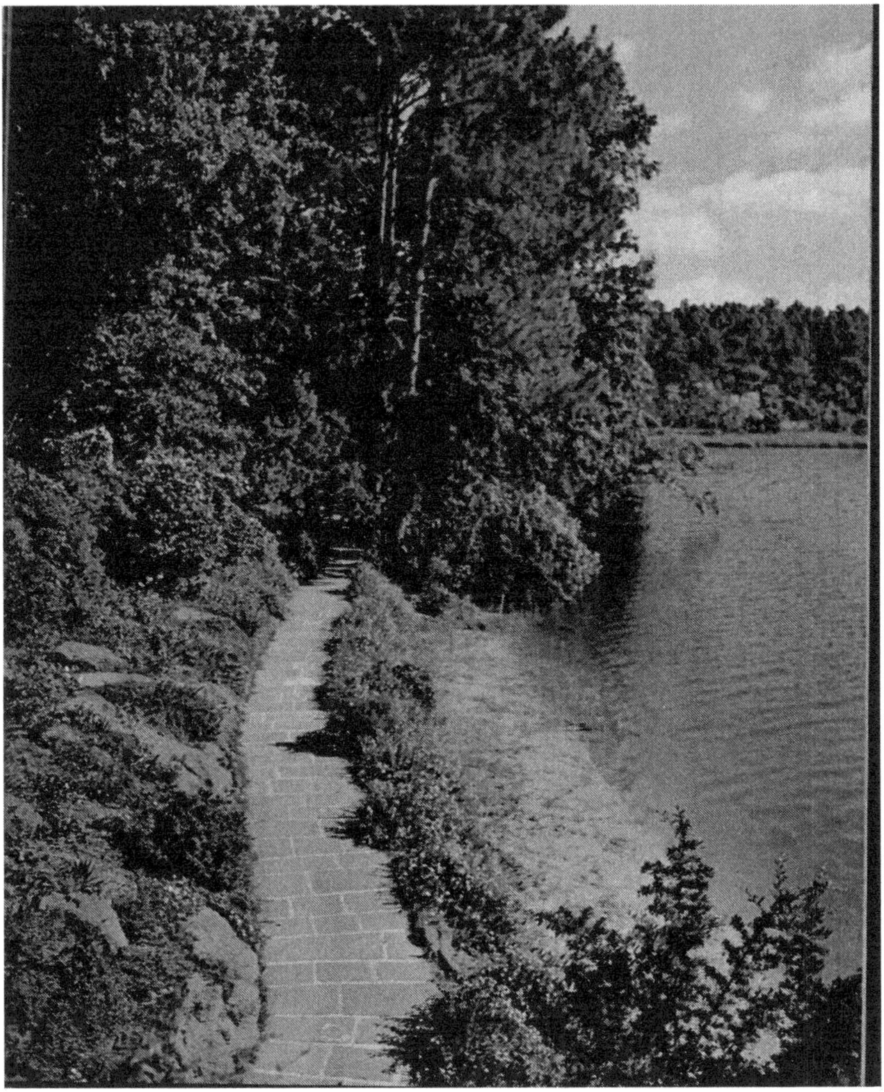

Small craft from nearby resorts bring guests to this boathouse on the river.

Golden allamandas bloom
profusely along the Isle-aux-Oies
during summer months;
the framing leaves are
on a sweetgum tree.

Through the years pictures of the Grotto have been used to epitomize the Gardens; in bloom are azaleas, cineraria, Easter lilies and chrysanthemums.

Another view of the Grotto in very early summer. Note the foliage of giant camellia bushes in the background.

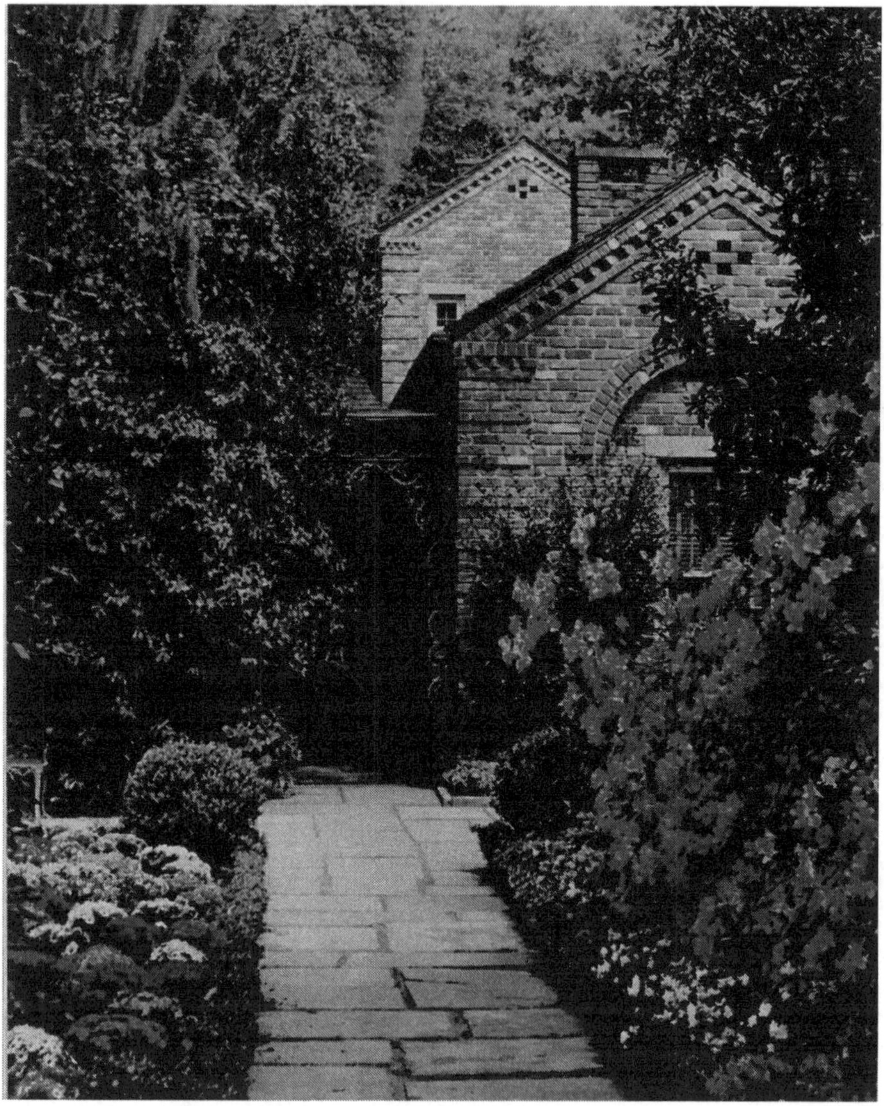

A view of the west wing of
The Bellingrath Home.
On the left are cineraria,
on the right azaleas.

Poinsettias and Belgian hybrid azaleas bring Christmas cheer to the granite monolith at the Gardens' center. The bronze plaques tell the story of the Gardens and their founders.

This bronze bas-relief of Rebecca at the Well
is surrounded by yellow calla lilies
and pink hydrangeas.

The more formal portions of
Bellingrath Gardens are designed so that
inspiring vistas may be seen in every direction.

Flower-banked fountains play in the shade of spreading trees that have seen Indian,
French, Spanish, British and American settlers in this part of the world.

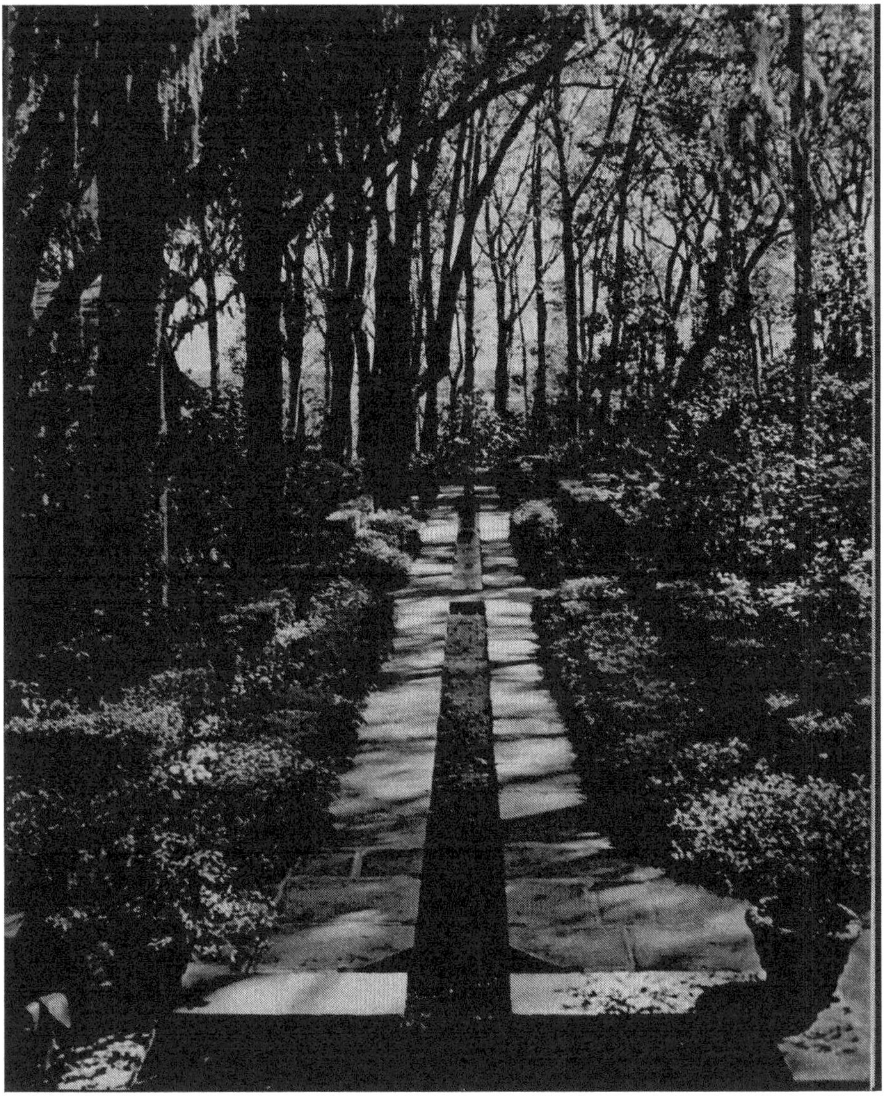

The waters of the Isle-aux-Oies may be seen through the trees.

Japanese Kurume azaleas, clipped hedges, and Indica azaleas border this pathway.

A small portion of the Great Lawn is glimpsed beyond beds of flaming salvia.

The lawn is surrounded on three sides by azaleas. The deeper pink are the Pride of Mobile variety.

A traditional Southern favorite crepe myrtle, is shown in this bed, bordered by foliage plants and annuals.

An exquisite Oriental magnolia blooms beneath a Southern magnolia.

The Rose Garden is in heavy bloom from April through December. Scores of varieties of roses are found here.

The lily ponds in the Rose Garden are surrounded by pansies, calendulas and other seasonal flowers.

The Peace rose is a favorite variety. Others include Etoile d'Holland and the dark red Night rose.

A closer view of the lily pond shows the delicate coloring of these fragile flowers. They are natives of the tropics.

The greenhouse at the entrance to the Rose Garden, which is in the form of the Rotary emblem, is filled with tropical and semi-tropical plants.

History

OF BELLINGRATH GARDENS

This early painting shows the rustic site of now fabulous Bellingrath Gardens, which began as a fishing lodge for Walter D. and Bessie Morse Bellingrath in 1917. It was Mrs. Bellingrath who first began planting azaleas in the woods around the lodge. So successful were her efforts that she and her husband soon became enthusiastic over the possibility of creating a wondrously beautiful garden from the forest around them.

In 1927, on a trip to Europe, the Bellingraths were enormously impressed by the great gardens they found there. They decided to call upon professional landscape architects to help them in their labor of love on the Isle-aux-Oies River. The aid of George B. Rogers, internationally known landscape designer and architect, was enlisted and the major aspects of the transformation were begun.

Not until 1932 were the Gardens first opened to the public. So overwhelming was the response to the Bellingraths invitation to come see their gardens that the highway patrol was called to help untangle the traffic snarl. To insure an appreciative audience and to help with the tremendous cost of upkeep, it was decided that an admission fee must be charged. This has been customary ever since.

The Gardens grew in size as additional acres of woodlands surrounding them were landscaped and planted. One of the interesting developments is the Camellia Arboretum, which is intended to provide the Camellia enthusiast an opportunity to compare the growing and flowering habits of the many, many varieties of this queen of all flowers. This is perhaps the most complete collection of its kind in the world.

Both Mr. and Mrs. Bellingrath had the satisfaction of seeing their fondest dream become a reality in the unsurpassed beauty of Bellingrath Gardens. Had they been buried in the Gardens instead of in the family plot in Magnolia Cemetery in Mobile, their epitaph might well have been copied from Sir Christopher Wren's in St. Paul's Cathedral, which he designed: "If you seek a monument, look about you."

The Bellingrath-Morse Foundation

The purpose of The Bellingrath-Morse Foundation is best explained in Mr. Bellingrath's own words as written in the preamble to the Deed of Trust creating the Foundation on February 1, 1950.

> "In the evening of our lives my beloved wife, Bessie Morse Bellingrath, and I found untold pleasure and happiness in the development of the Gardens which bear our name. During the past decade thousands of our fellow citizens have enjoyed the rare and lovely spectacle which nature, with our help, has provided in this 'Charm Spot of the Deep South.' The inspiration which we received as we carried on our work of developing the Gardens and the pleasant and appreciative reaction of the many visitors to the Gardens resulted in plans for the perpetuation of this beauty, so that those who come after us may visit the Gardens and enjoy them. In working out our plans, it occurred to us that the operation of the Gardens could be carried on in a way that would continue their existence and yet fulfill another worthy objective of ours. To this end, I am providing herein that the income from the operations of the Gardens be devoted to the intellectual and religious upbuilding of young men and women, as well as to foster and perpetuate those Christian values which were recognized by our forefathers as essential for the building of a great nation."

Mr. Bellingrath died on August 8, 1955, leaving the bulk of his estate to the Foundation. Today the Gardens are under the administration of the six trustees of The Bellingrath-Morse Foundation. The Corporate Trustee is the First National Bank of Mobile, of which Mr. Bellingrath was a director. The five individual trustees are men who, over a period of time, were closely associated with Mr. Bellingrath.

With the establishment of The Bellingrath-Morse Foundation, Southwestern at Memphis, Huntingdon College at Montgomery, and Stillman College (for Negroes) at Tuscaloosa became beneficiaries along with the Central Presbyterian Church of Mobile (as a perpetual memorial to his parents), and the St. Francis Street Methodist Church of Mobile (as a perpetual memorial to his wife's parents).

It is incumbent upon the colleges in order to qualify as a beneficiary that each student shall be required to take a course in Bible training. Thus, through the Foundation, Mr. Bellingrath is carrying out a long fostered and cherished plan of perpetuating world famed Bellingrath Gardens and, at the same time, providing long term benefits to the colleges which carry on in the Christian tradition.

MAP OF BELLINGRATH GARDENS

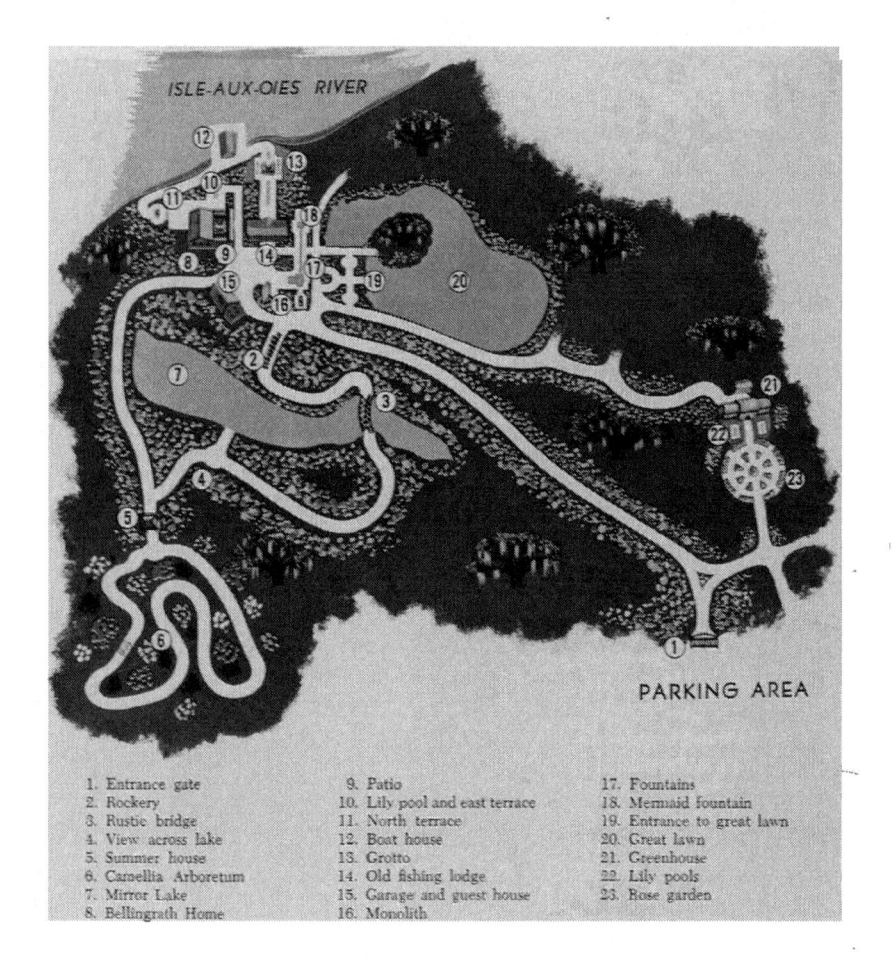

1. Entrance gate
2. Rockery
3. Rustic bridge
4. View across lake
5. Summer house
6. Camellia Arboretum
7. Mirror Lake
8. Bellingrath Home
9. Patio
10. Lily pool and east terrace
11. North terrace
12. Boat house
13. Grotto
14. Old fishing lodge
15. Garage and guest house
16. Monolith
17. Fountains
18. Mermaid fountain
19. Entrance to great lawn
20. Great lawn
21. Greenhouse
22. Lily pools
23. Rose garden

Photography by: Bill Shrout, William W. Lavendar, Fred. W. Holder

COLOR LITHOGRAPHY BY LITHO-KROME COMPANY, COLUMBUS, GEORGIA

Printed in the USA
CPSIA information can be obtained
at www.ICGtesting.com
LVHW011025150823
755292LV00006B/67